Living or Not living

 Paste each sticker in the correct group.

Living

grass

sunflower

dog

human

Not living

notebook

ice cream

rock

toy

Chapter 1
Life Science

Plants and Animals

Facts Living things can be classified into two types, plants and animals.

Circle the animals in the picture.

| eagle | duck | bear | deer | squirrel |

Chapter 1
Life Science

Plant Parts

Facts Plants have parts such as flowers, leaves, stems, and roots. Each part has a role.

 Paste each sticker in the correct place.

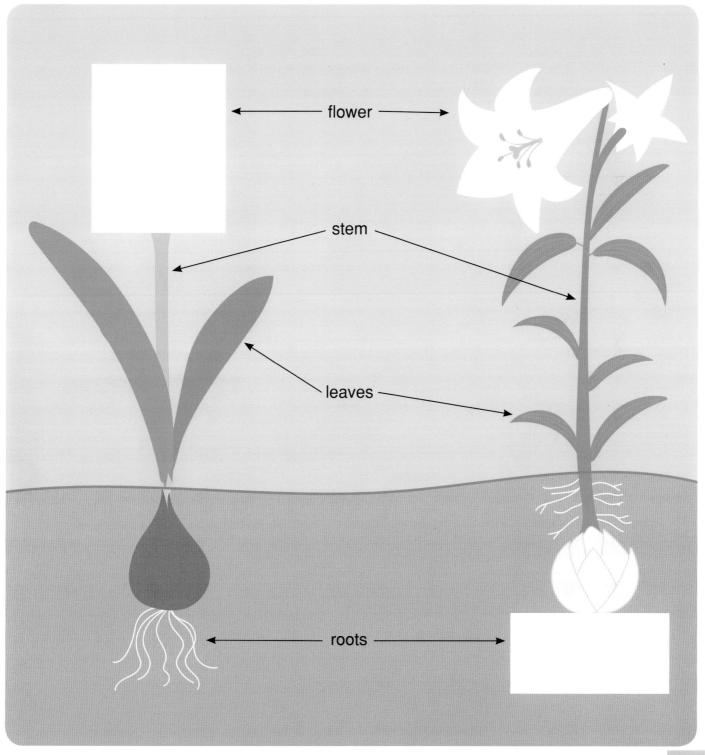

Chapter 1
Life Science

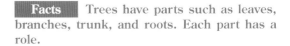

Facts   Trees have parts such as leaves, branches, trunk, and roots. Each part has a role.

Tree Parts

 Paste each sticker in the correct place.

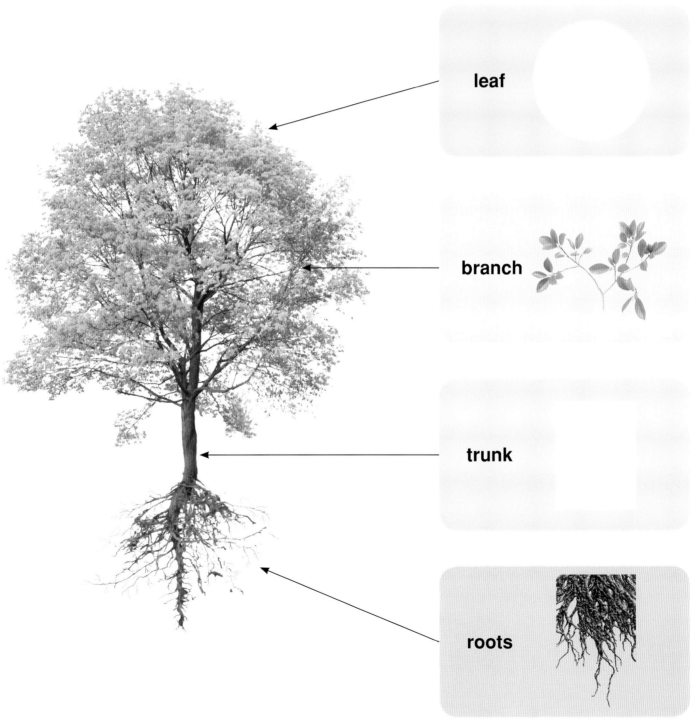

leaf

branch

trunk

roots

Life Science

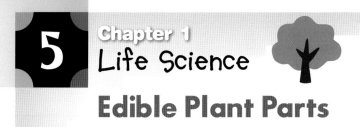

Facts Some plants grow parts that can be eaten. Vegetables and fruits are parts of plants that we can eat.

Edible Plant Parts

 Paste each sticker in the correct group.

We can eat fruit.

apple

We can eat seeds.

green peas

We can eat roots.

carrot

We can eat leaves.

lettuce

Life Cycle of Plants

Facts Plants change and grow. Most plants begin as seeds and grow to be trees and flowers.

Draw a line through the maze from the arrow (➡) to the star (★). See how the watermelon grows.

Life Science

Life Cycle of Animals

Facts Animals change and grow. They have different life stages.

 Paste each sticker in the correct place.

butterfly

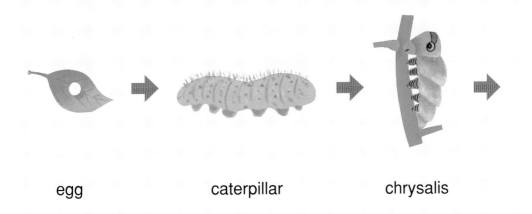

egg   caterpillar   chrysalis

human

baby   toddler   teenager

Life Science

Classification of Animals

Facts There are different kinds of animals such as mammals, birds, fish, reptiles, and amphibians.

 Paste each sticker in the correct group.

mammal

monkey

cat

human

bird

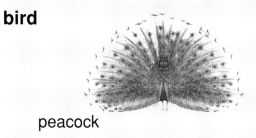

peacock

chicken

fish

salmon   tuna   shark

reptile

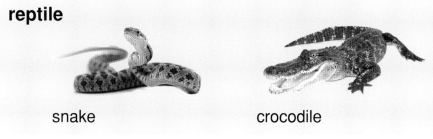

snake   crocodile

amphibian

frog   newt   salamander

Life Science

Animal Habitats

Paste each sticker in the correct group of their habitat.

forest

fox

pond

frog

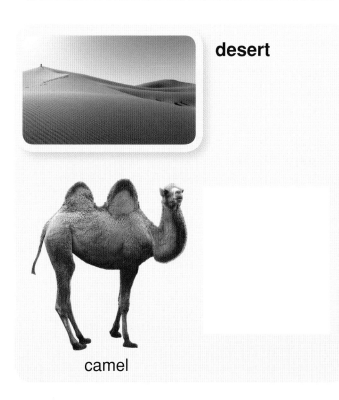

desert

camel

cave

bear

Chapter 1
Life Science

Facts Some animals have body parts for protection such as fur, feathers, scales, or shells.

Animal Parts for Protection

 Draw a line to match each animal with a body part for protection.

sheep

 shell

swan

 feathers

lizard

 fur

snail

 scales

Chapter 1
Life Science

Face Parts

Facts The face has parts such as eyes, nose, ears, and mouth.

 Paste each sticker in the correct place.

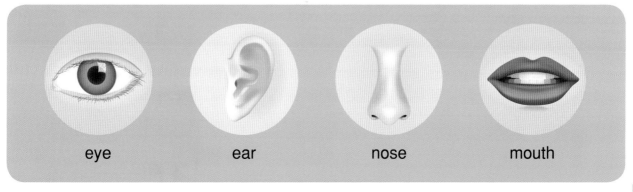

| cheyye | ear | nose | mouth |

Life Science

Body Parts

Facts The human body has many parts such as head, shoulder, hand, knee, and foot. Each part has a role.

Draw a line to match each word to the correct body part.

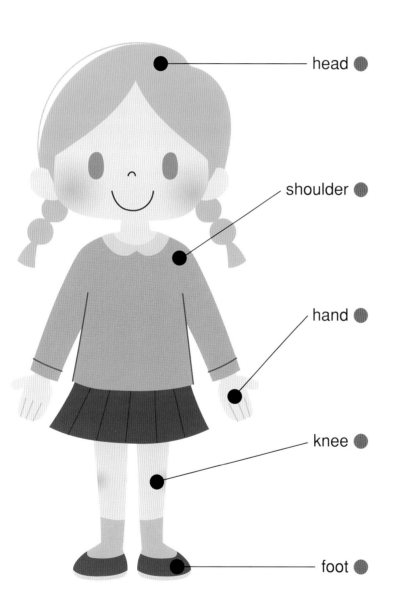

head ●

shoulder ●

hand ●

knee ●

foot ●

13 Chapter 1
Life Science
Organs

Facts The body has organs that do different jobs to keep living things alive.

Draw a line to match each word to the correct organ.

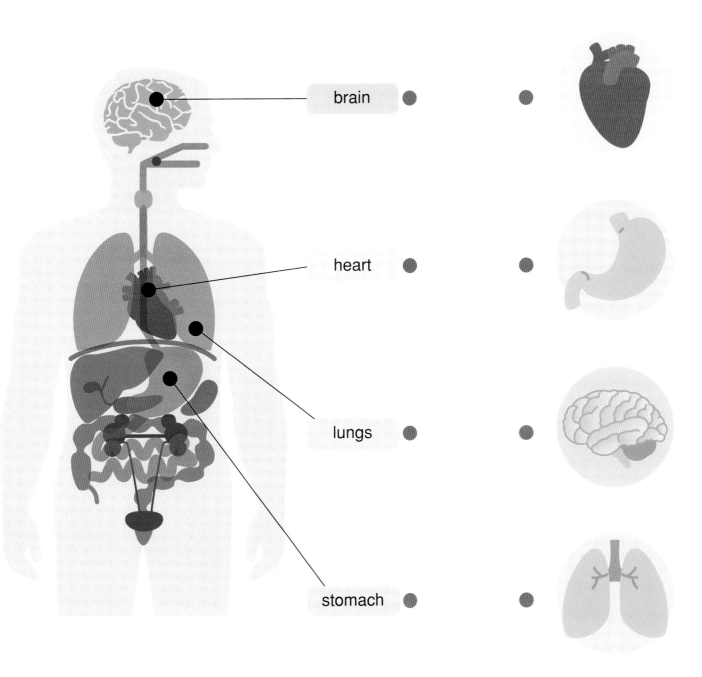

brain

heart

lungs

stomach

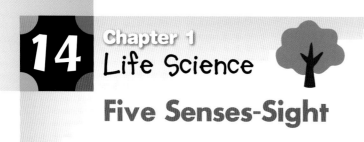

Chapter 1
Life Science

Five Senses-Sight

Facts Sight is one of the five senses. We see various things with our eyes.

 Paste each sticker in the missing eyes.

dog

owl

human

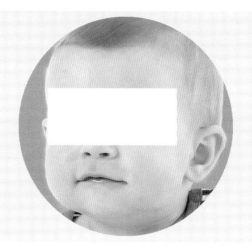

Chapter 1
Life Science

Five Senses-Hearing

Facts Hearing is one of the five senses. We listen to various sounds with our ears.

Color the things you can hear.

You can hear.

bell   guitar

drum

You can NOT hear.

bag   banana   notebook

mushroom   hat

Life Science

Five Senses-Smell

 Draw a line to match each nose to the correct animal.

gorilla

human

pig

elephant

Chapter 1
Life Science

Five Senses-Taste

Facts Taste is one of the five senses. We taste various foods with our tongues. There are parts of the tongue particularly sensitive to each taste.

Paste each sticker in the correct place.

sugar

ice cream

salt

french fries

lemon

vinegar

coffee

kale

Five Senses-Touch

Facts Touch is one of the five senses. By touching objects with our hands we can feel if they are hard, soft, hot, or cold.

Color the things that feel soft.

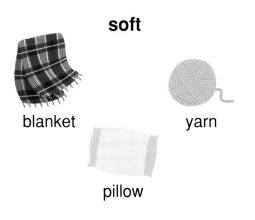

soft

blanket   yarn

pillow

hard

scissors   car   chair

juice bottle   mobile phone

Chapter 1
Life Science

What Animals Eat

Facts Animals need food to live. Some animals eat plants, others eat meat from other animals, and some animals eat both plants and meat.

🔍 Look at the picture. What does each animal eat? Circle what they eat.

bear   crayon   fish

bird   worm   soda

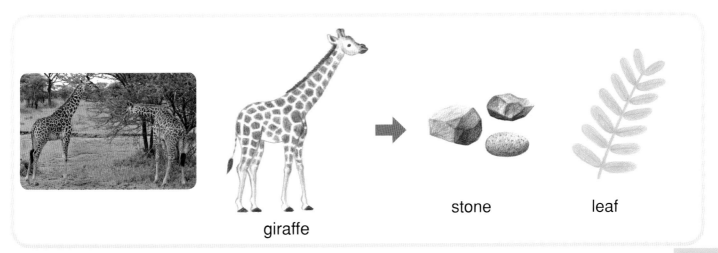

giraffe   stone   leaf

Chapter 1
Life Science

Facts A food chain exists in nature. Food chains show what animals eat to get energy. Animals get energy from eating plants or other animals.

Food Chain

 Paste each sticker in the correct group.

carnivore

lion   snake

eat

herbivore / omnivore

zebra   sparrow

eat

plants

fallen leaves   nuts

Microscope

Facts Using a microscope, you can observe small objects that you cannot usually see.

Paste each sticker in the correct place.

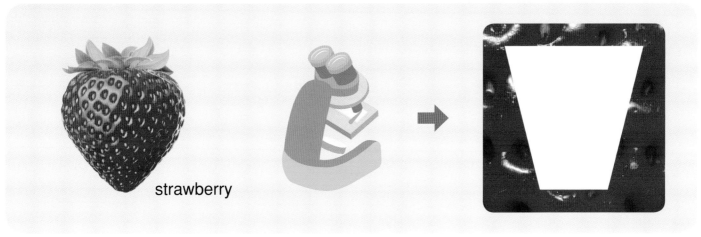

strawberry

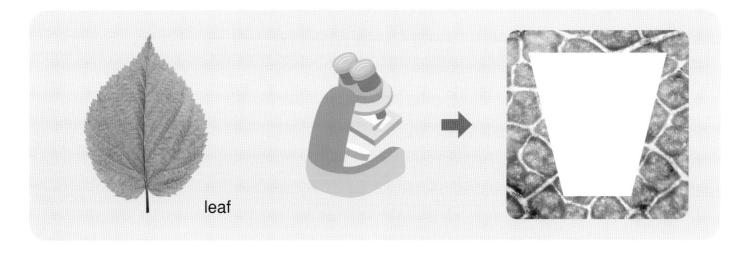

leaf

snow

Chapter 2
Earth Science

Land

Facts There are different types of land on the earth, such as mountains, deserts, plains, and islands.

 Paste each sticker in the correct place.

plain

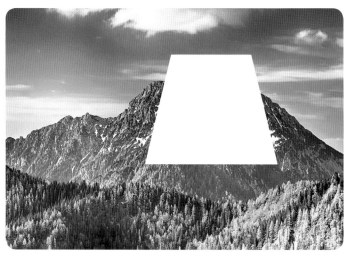

mountain

desert

island

Chapter 2
Earth Science

Water

Facts There are different water formations on Earth such as lakes, rivers, and oceans. Fresh water can be found in lakes and rivers. Ocean water is salty.

 Color the water blue.

lake

river

ocean

Earth Science

Rocks

Facts You can see rocks of various sizes and shapes. As the big rocks upstream travel downstream they collide with other rocks, gradually rounding and becoming smaller.

 Paste each sticker in the correct place.

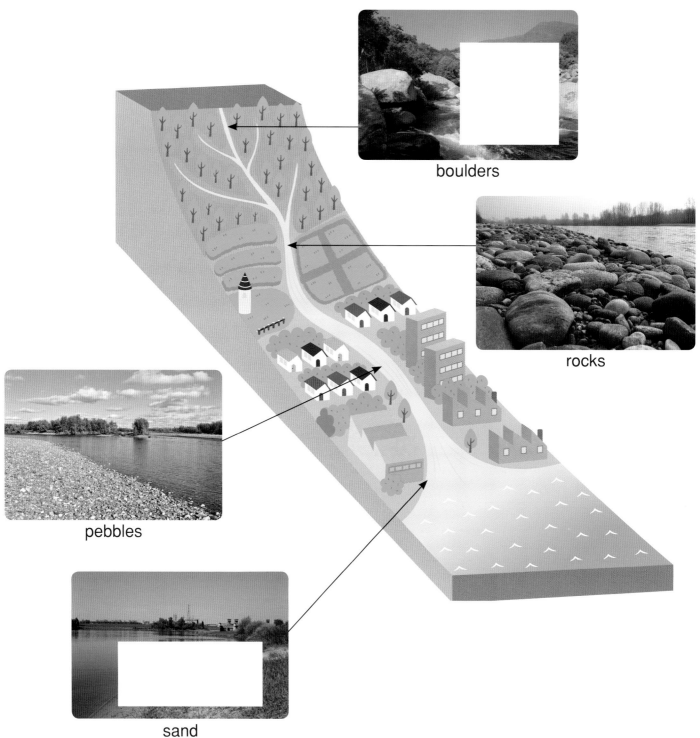

boulders

rocks

pebbles

sand

Facts There are natural materials (such as wood and stone) and things made by human hand by using natural materials.

Natural or Man Made

 Paste each sticker in the correct group.

Natural

tree

rock

soil

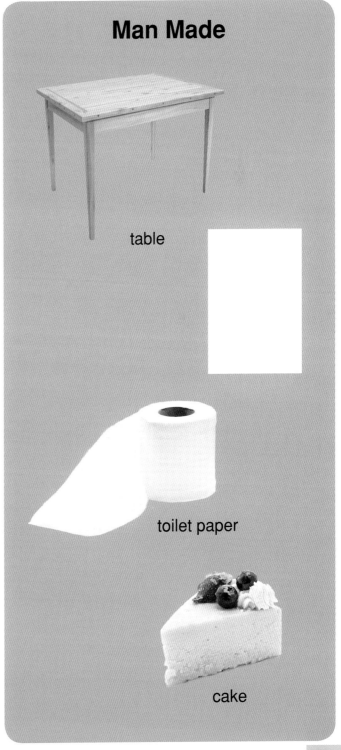

Man Made

table

toilet paper

cake

Moon Phases

Facts The moon changes shape during each month. Different moon shapes are called moon phases. You can see the same phases every month.

 Paste each sticker in the correct place.

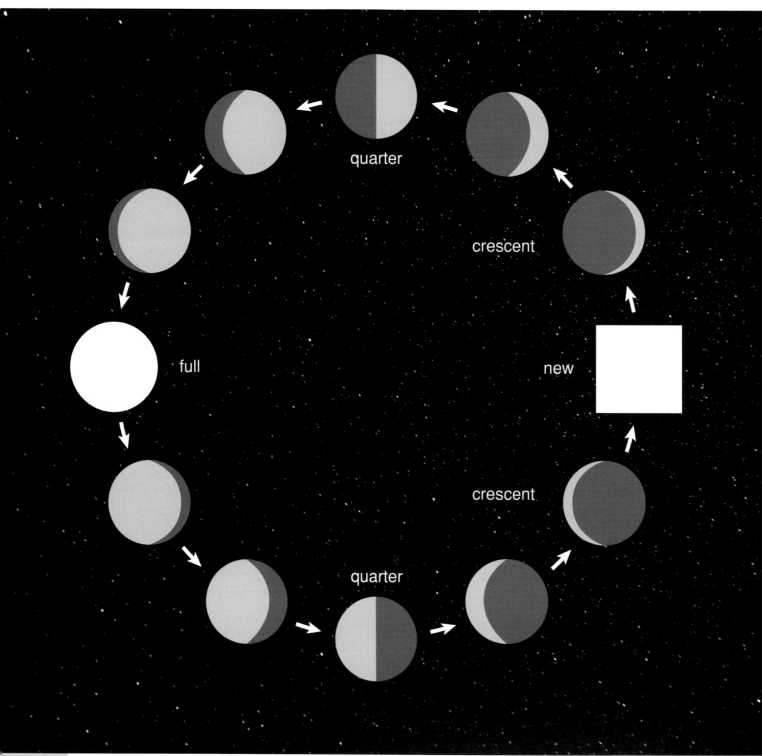

quarter

crescent

full

new

crescent

quarter

Chapter 2
Earth Science

Sun, Earth, and Moon

Facts The sun is a star. It looks bigger than other stars because it is close to the Earth. The Earth moves around the sun. The moon moves around the Earth.

Paste moon and sun stickers in the picture below.

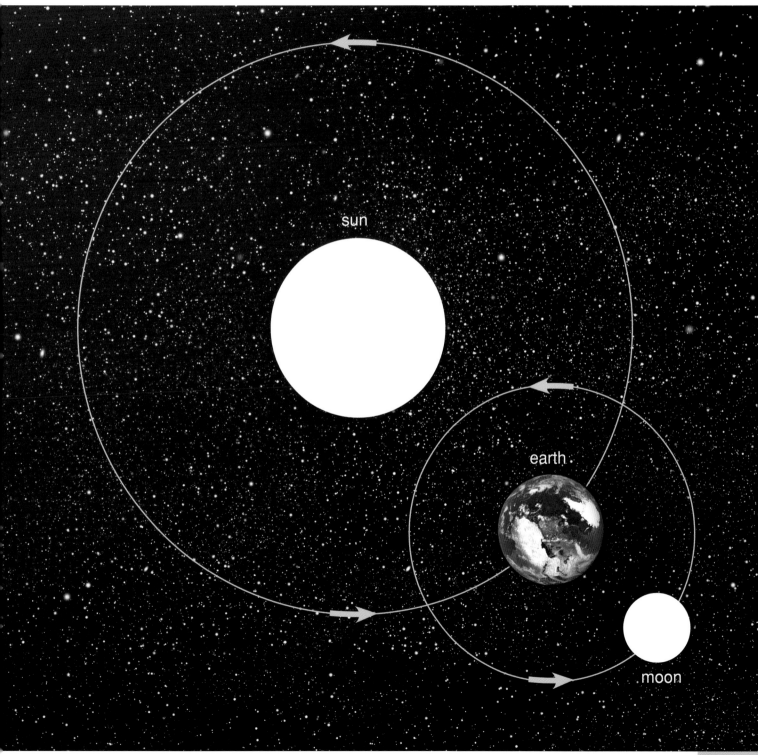

 **28** **Chapter 2**
Earth Science

Weather

Draw a line to match each weather situation to the correct word.

 ● rainy

● snowy

 ● sunny

 ● windy

Earth Science

Facts Clouds can be made of water droplets or ice crystals. There are different types of clouds such as cirrus, stratus, cumulus, and cumulonimbus.

Clouds

Look at the pictures of the types of clouds. Then draw clouds on the picture of the sky below.

cirrus

stratus

cumulus

cumulonimbus

Chapter 2
Earth Science

Rainbow

Facts A rainbow is a colored arch that appears when the sun's rays shine through rain drops. When the sun appears after the rain stops, a beautiful rainbow can be seen.

 Color the rainbow.

red
orange
yellow
green
blue
indigo
violet

Earth Science

Facts There are various tools for measuring weather conditions such as temperature, amount of rain, and strength of wind.

Weather Tools

Paste each sticker in the correct place.

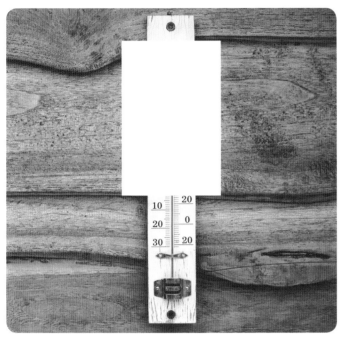

thermometer

rain gauge

anemometer

weather vane

Earth Science

Seasons

Facts The Earth has four seasons. They are spring, summer, fall, and winter. There is an order to the four seasons.

 Paste each sticker in the correct place.

Chapter 2
Earth Science

Recycling

Facts We can recycle objects and use them to make new things. When we recycle, we sort objects by what they are made of. We can recycle paper, plastic, and glass.

 Paste each sticker in the correct place.

paper → books

plastic → clothing

glass → new bottle

Physical Science

Shape

Facts Shape is one way to describe things. You can find many examples of 2D and 3D shapes in your daily life.

 Draw a line to match two objects that are similar in shape.

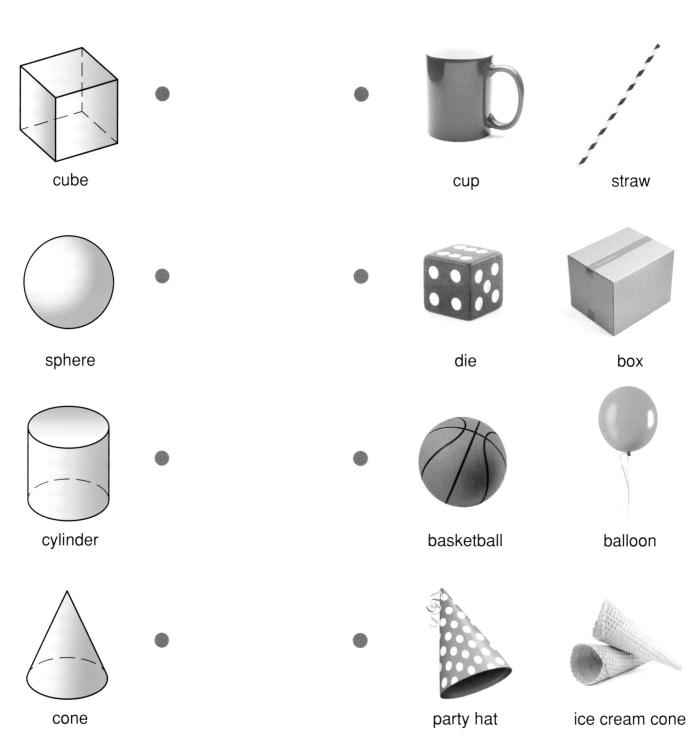

cube

cup

straw

sphere

die

box

cylinder

basketball

balloon

cone

party hat

ice cream cone

Physical Science

Size

Facts Size is one way to describe things. You can compare the size of multiple objects and rearrange them in ascending order or descending order.

 Paste each sticker in order from largest to smallest.

melon   pepper   strawberry

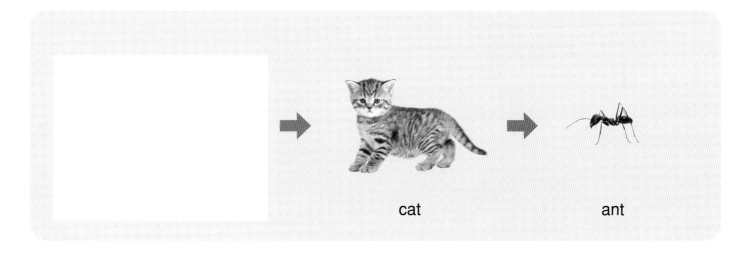

cat   ant

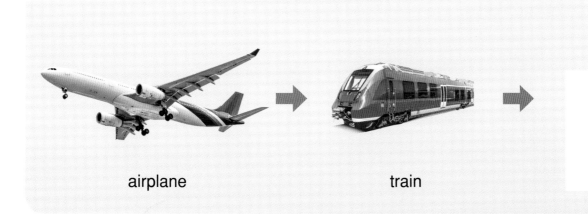

airplane   train

Chapter 3
Physical Science

Color

Facts Color is one way to describe things. Many people have a favorite color. Do you have one?

Use the code to color the picture.

1 orange   2 green   3 yellow   4 red   5 blue

Physical Science

Texture

Facts Texture is one way to describe things. There are different textures such as soft, rough, and smooth.

Draw a line to match the object that has a similar texture.

smooth

bumpy

fluffy

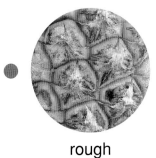

rough

Weight

Facts Weight is one way to describe things. You can compare the weight of multiple objects and rearrange them in ascending order or descending order.

Paste each sticker in order from the heaviest to lightest.

hippopotamus   rabbit   sparrow

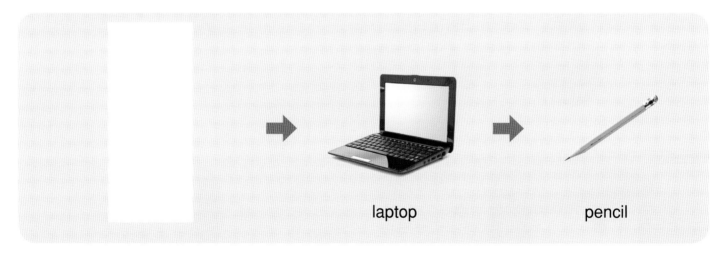

laptop   pencil

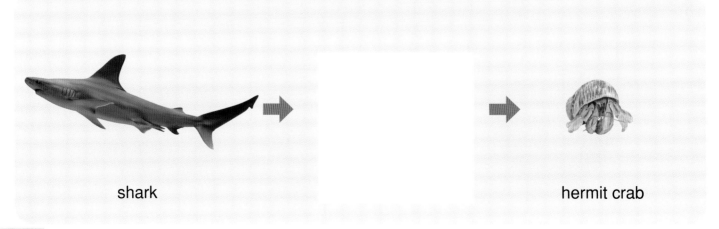

shark   hermit crab

Chapter 3
Physical Science

Sink or Float

Facts Objects float or sink in water. Some objects will float on top of the water and others will sink below the water.

39

Paste each sticker in the correct place.

leaf

sponge

green pepper

coin

scissors

rock

Position

Facts There are effective words when explaining the position of an object.

 Draw a line to match the word with the same position.

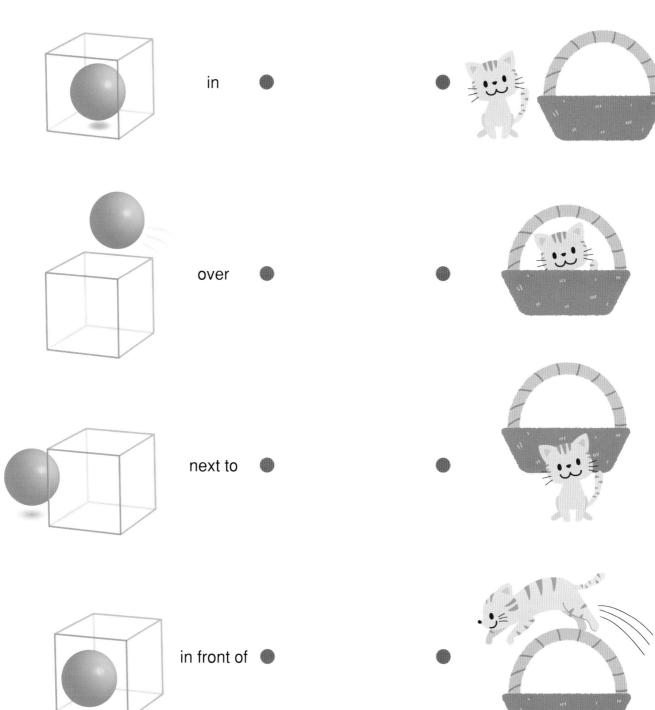

in

over

next to

in front of

Chapter 3
Chapter 3
Physical Science
Moving Things

Facts There are objects that can and cannot move by themselves.

Paste each sticker in the correct group.

Move by Themselves

dog

crab

pig

Do Not Move by Themselves

daisy

tomato

lamp

Chapter 3
Physical Science

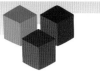

Facts Objects can move at various speeds such as fast and slow.

Fast or Slow

Paste each sticker in order from fastest to slowest.

cheetah   chicken   snail

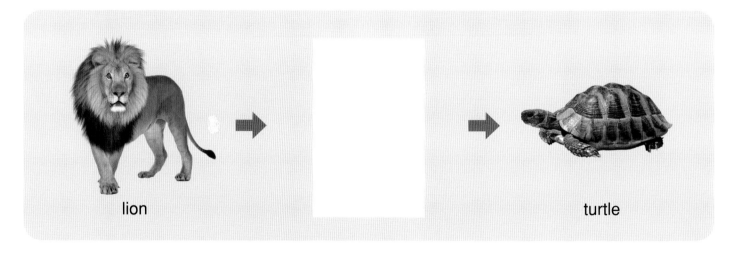

lion   turtle

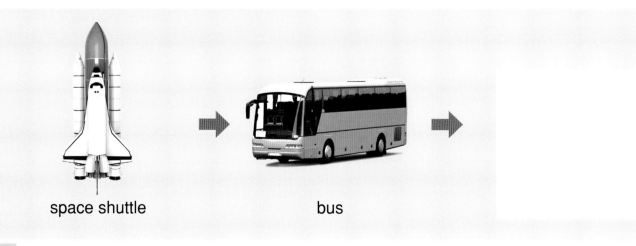

space shuttle   bus

Chapter 3
Physical Science

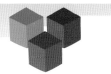

Facts Objects can move in various directions such as back and forth, up and down, zigzag, and in a circle.

The Way Things Move

 Draw a line to match the picture with the same motion.

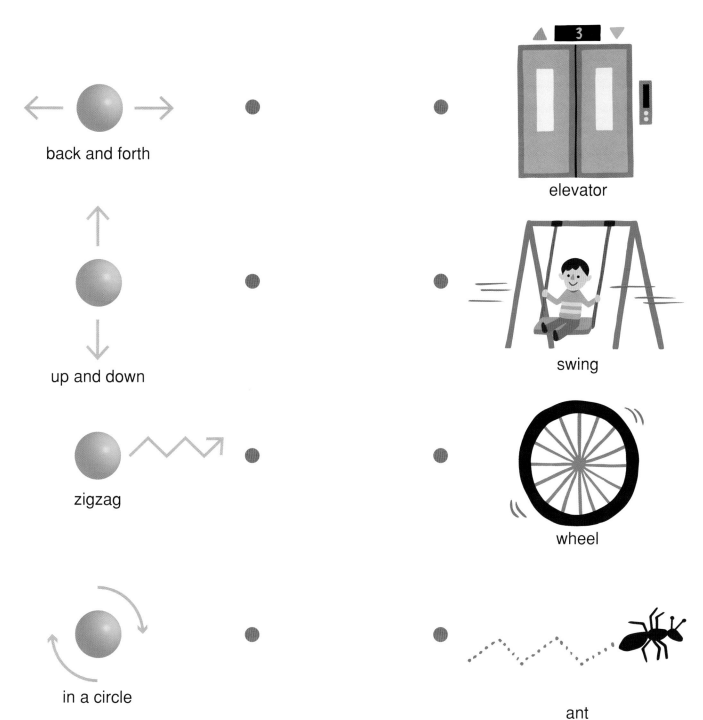

back and forth

up and down

zigzag

in a circle

elevator

swing

wheel

ant

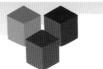

Facts Push and pull are types of forces that make objects move.

Forces

 Paste each sticker in the correct group.

Push	Pull

shopping cart

swing

button

rope

toy wagon

zip

Physical Science

Solid, Liquid, or Gas

 Draw a line to match a solid, liquid, or gas.

solid

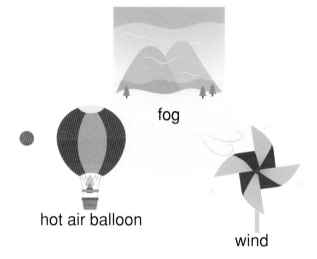

fog

hot air balloon

wind

liquid

hamburger

toothbrush

telephone

gas

gasoline

paints

milk

Chapter 3
Physical Science

Facts Objects can change substance from solid to liquid or liquid to gas when they are heated.

Solid to Liquid

Draw a line through the maze from the arrow (➡) to the star (★). See how objects change when they are heated.

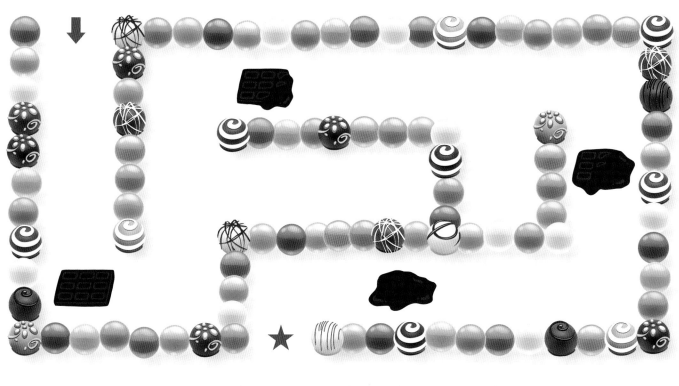

Chapter 3
Physical Science

Magnets

🔍 Look at the picture of the magnet. Then circle all the objects that will stick to the magnet.

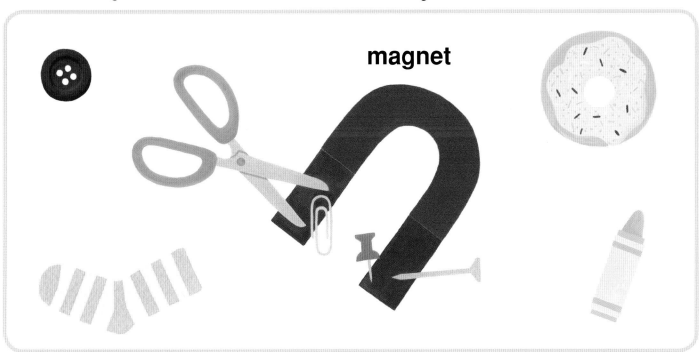

magnet

socks	paper clip	nail	donut
scissors	button	crayon	pin

Physical Science

Measuring Tools

Facts There are useful tools to measure the length, weight, and quantity of objects.

 Paste each sticker in the correct place.

page 1

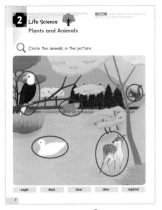

page 2

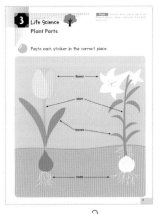

page 3

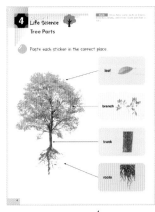

page 4

page 5

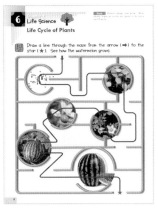

page 6

page 7

page 8

page 9

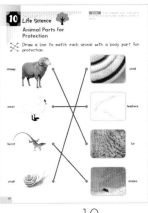

page 10

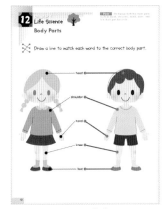

page 11

page 12

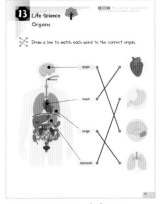

page 13

page 14

page 15

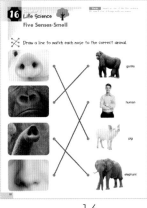

page 16

page 17

page 18

page 19

page 20

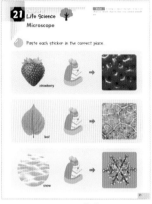
page 21

page 22

page 23

page 24

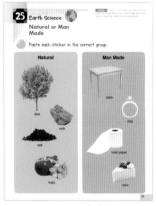
page 25

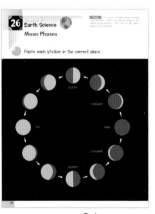
page 26

page 27

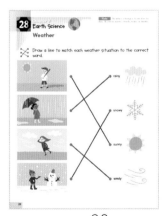
page 28

page 29

page 30

page 31

page 32

page 33

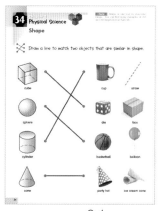

page 34

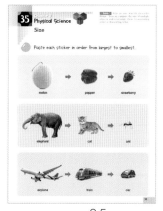

page 35

page 36

page 37

page 38

page 39

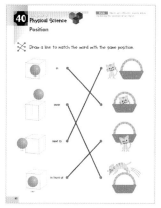

page 40

page 41

page 42

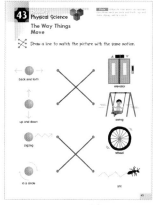

page 43

page 44

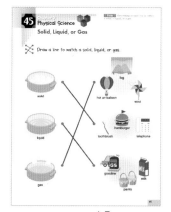

page 45

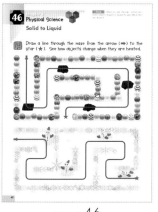

page 46

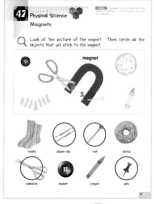

page 47

page 48

Certificate
of
Achievement

is hereby congratulated on completing

Kumon Sticker Activity Books

SCIENCE

Presented on _____ , 20 _____

_____

Parent or Guardian

KUMON

tree

fish

To be used in **1**

stuffed animal

car

To be used in **3**

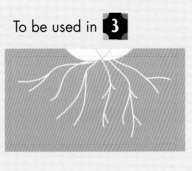

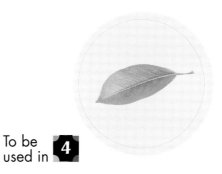
To be used in **4**

peach

butterfly

adult

To be used in **7**

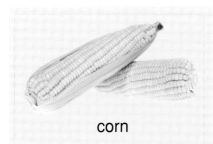
corn

To be used in **5**
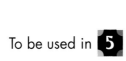

eagle

radish

spinach

To be used in **8**

turtle

squirrel

duck

lizard

bat

To be used in **9**

To be used in **11**

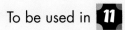
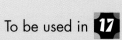

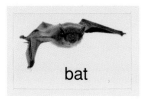

To be used in **14**

eagle

chipmunk

To be used in **17**

sweet

salty

To be used in **20**

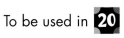
grass

sour

bitter

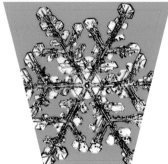

To be used in **21**

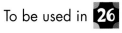

To be used in **26**

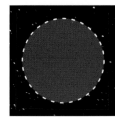

To be used in **22**

To be used in **25**

fruits

ring

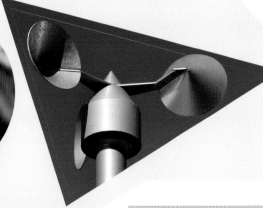

To be used in **24**

To be used **27**

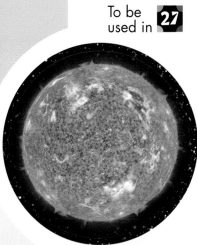

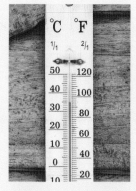

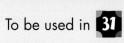
To be used in **31**

To be used in **32**

toilet paper

hanger

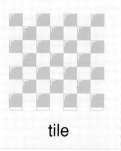
tile

To be used in 33

elephant

refrigerator

squid

To be used in 41

To be used in 35

car

giraffe

soccer ball

To be used in 38

To be used in 39

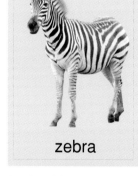
pencil

carrot

zebra

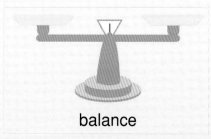
bicycle

To be used in 42

stroller

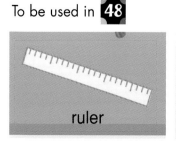
fishing

To be used in 44

stop watch

balance

To be used in 48

ruler

measuring cup